a Guide to church inspection and repair

a Guide to church inspection and repair

2nd edition

CHURCH HOUSE PUBLISHING

Church House Publishing
Church House
Great Smith Street
London
SW1P 3NZ

ISBN 0 7151 7568 8

First published 1980
Second edition 1995
Reset 2001

Copyright assigned to the
Archbishops' Council 1999.

Printed by Halstan & Co. Ltd
Amersham, Bucks

All rights reserved. No part of
this publication may be reproduced
or stored or transmitted by any
means or in any form, electronic or
mechanical, including photocopying,
recording, or any information storage
and retrieval system without written
permission which hould be sought
from the Copyright and Contracts
Administrator, The Archbishops'
Council, Church House, Great
Smith Street, London SW1P 3NZ
(Tel: 020 7898 1557;
Fax: 020 7898 144;
copyright@c-of-e.org.uk).

Contents

Introduction	vii
The inspection of church buildings	1
The appointment of a professional adviser	3
Fees for quinquennial inspections	5
Preparing for the inspection	6
Keys	6
Ladders	6
Inventory, log book and other information	6
Security	7
Bells	7
The inspection and report	8
Putting the work in hand	9
Planning	9
Funding the work	10
Seeking approval	10
Faculty jurisdiction	11
Permissions from secular authorities	12
Consulting other people and bodies	12
Selecting the contractor	13
Insurance	14
Value added tax	15
appendix 1 Model diocesan scheme	16
appendix 2 Draft letter of appointment	17
appendix 3 Outline quinquennial inspection report	20
appendix 4 Advice to DACs on assessing potential professional advisers	29

Contents

appendix 5 Churchwardens' responsibilities under the Care of
Churches and Ecclesiastical Jurisdiction Measure 1991   34

appendix 6 A guide to the routine maintenance and inspection
of church poverty   36

appendix 7 A guide to getting the work started   39

appendix 8 A guide to the faculty procedure   40

appendix 9 Further reading   41

appendix 10 Useful addresses   42

Introduction

The aim of this booklet is to give information and guidance on the systems established by the Church of England for the inspection of churches and the undertaking of suitable repairs, in line with statutory requirements and in particular the Care of Churches and Ecclesiastical Jurisdiction Measure 1991.

The survival of so many English churches from the Middle Ages and later is due principally to the care of successive generations of parishioners who have invested energy and money in their local church because it is the focus of their prayer and worship, and a physical witness to the glory of God. The attraction of a church extends beyond its regular congregation to the wider community, especially where the church is attractive and welcoming; unloved and poorly kept it conveys a negative message, encouraging theft and vandalism.

The saying "a stitch in time saves nine" illustrates a fundamental precept for the repair of old buildings. Prompt attention to defects and a regular planned programme of maintenance and minor repairs will reduce the need for sudden urgent works.

The first guide to church inspection and repair appeared in 1970, and was a practical working document specifically aimed at architects involved in or aspiring to historic building repairs. This latest edition is more than an update of its predecessors. It contains new guidelines taking into account the effects of the Care of Churches and Ecclesiastical Jurisdiction Measure 1991. It refers to documents which have been recently issued, especially the Code of Practice, published by Church House Publishing, the Faculty Jurisdiction Rules published by HMSO, and the list of sources of grant aid produced by the Council for the Care of Churches which is available free on enquiry.

In 1970 it was the Ecclesiastical Architects' and Surveyors' Association in collaboration with the then Council for the Care of Churches, and in particular Miss Judith Scott, who initiated the working document. The Council is indebted to the co-operation

Church inspection and repair

provided by the Ecclesiastical Architects' and Surveyors' Association, and also to the Royal Institution of Chartered Surveyors and the Royal Institute of British Architects, for the time and trouble taken to produce what we hope will be a worthy successor.

The early drafts of this booklet were produced by a working party consisting of David Barclay and Robert Tolley representing the Royal Institute of British Architects, Geoffrey Claridge representing the Ecclesiastical Architects' and Surveyors' Association, Tony Redman representing the Royal Institution of Chartered Surveyors and Jonathan Goodchild representing the Council for the Care of Churches. Thanks should also be expressed to Thomas Cocke, Richard Halsey, Hester Agate and Ingrid Slaughter who read and corrected the drafts and to Ann Owens of the Whitworth Co-Partnership who typed, corrected and re-typed the document. Thanks especially to Tony Redman who collated the working group's thoughts and masterminded the final editing with Thomas Cocke.

This working document will be of use not only to architects and chartered building surveyors involved in the quinquennial inspection of churches under the Inspection of Churches Measure of the Church of England, but to all who have some responsibility for the care and maintenance of churches, including diocesan and other client bodies. Whilst produced specifically for the Church of England, there will be material here of interest to those other Churches for whom, under the revised arrangements for ecclesiastical exemption issued by the Department of National Heritage, the need to provide an approved system for the administration of building repairs to listed buildings has recently become an issue of some concern.

✠ Colin Hulme
Chairman
Council for the Care of Churches

February 1995

The inspection of church buildings

In 1955 the Church of England gave a lead to guardians of historic buildings by passing a law requiring a regular inspection every five years of all parish churches.

The legislation, as now amended by the Care of Churches and Ecclesiastical Jurisdiction Measure 1991, requires diocesan synods to establish a scheme for the inspection of every church in the diocese at least once every five years. This includes all parish churches, and also other consecrated chapels and churches and buildings licensed for worship by the diocesan bishop, subject to certain specific exclusions. The scheme must provide for:

a) The establishment of a fund to pay for inspections by means of contributions from parochial, diocesan or other sources;

b) The appointment of architects or chartered building surveyors approved by the Diocesan Advisory Committee for the Care of Churches (the DAC) to inspect each church in the diocese and to make a report;

c) The sending of a copy of the report to the archdeacon concerned, the parochial church council, the incumbent and the secretary of the DAC.

The 1991 Measure sets out for the first time some specific areas for the report to cover, in addition to the structure of the building. These are:

a) Any movable article in the church which the archdeacon concerned, after consultation with the DAC, requires the person carrying out the inspection to treat as being: of outstanding architectural, artistic, historical or archaeological value; or of significant monetary value; or at special risk of being stolen or damaged;

The person carrying out the inspection must also include other articles which they consider fall within any of the above categories, and may include further articles where that seems sensible.

Church inspection and repair

b) Any ruin in the churchyard (open or closed) jointly designated by the Council for British Archaeology and the Royal Commission on the Historical Monuments of England as being of outstanding architectural, artistic, historical or archaeological value;

Information on whether a ruin is designated may be available from these two bodies.

c) Any tree in the churchyard (open or closed) in respect of which a tree preservation order is in force. Details of tree preservation orders are available from the local authority.

A draft scheme for the inspection of churches recommended by the Council for the Care of Churches (the 'CCC') to diocesan synods is set out in Appendix 1.

The appointment of a professional advisor

In the remainder of this publication, architects and surveyors are referred to as the 'professional adviser', and the parochial church council 'PCC'.

In most diocesan schemes for the inspection of churches, the profess-ional adviser is appointed by the PCC with the agreement of the DAC. Most DACs operate an 'approved list' of professional advisers, while others consider each appointment individually. Guidelines to help DACs in assessing applicants are set out in Appendix 4. The relationship between a PCC and its professional adviser should ideally be a close and continuing one, enabling them to play a key role in the care and development of the church. The adviser will undertake the inspection, advise the PCC on how to implement the recommendations, and normally oversee the programme of repairs. The professional adviser should also be requested to advise the PCC on any fault that has developed in the building between inspections.

If the professional adviser retires, or if the PCC wishes to appoint a replacement for any other reason, the PCC should discuss possible candidates with the DAC secretary. The DAC's approval of an appointment is needed to ensure that a person is appointed who has appropriate knowledge and experience relative to the type, size and age of the building. In professions as complex as architecture and surveying, training and experience vary and it is important to ensure that these match the problems of looking after the particular building concerned. The parish may also have its own suggestion for a professional adviser, perhaps a member of the church or someone living in the community. However, this will not necessarily be the right solution: the candidate's experience might not be suitable, and a business relationship with such a person can lead to problems.

Church inspection and repair

The DAC will also have in mind the importance of introducing younger people into the field of church care and conservation, to ensure continuity of experience for the future.

The appointment is always a personal one although the adviser may be a member of a professional firm with others assisting.

Once the new professional adviser has been chosen by the PCC and approved by the DAC, the PCC secretary and the professional adviser should confirm the appointment in writing. A draft letter for this purpose is set out in Appendix 2.

Fees for quinquennial inspections

In some dioceses, the negotiation and payment of inspection fees are left to the PCC. In others, the fee is set and paid direct by the Diocesan Board of Finance. Whatever the method of payment, the CCC recommends that fees are calculated on the time needed to inspect the church and write the report. An approximate guide to the number of hours that may be needed to inspect churches of different categories is as follows:

simple churches	5 – 8 hours
average churches	8 – 10 hours
complex churches	10 – 15 hours
churches of unusual complexity or scale	to be negotiated on a case by case basis.

More time will be needed by a new professional adviser, to become familiar with the church and the previous work to it, and for a first inspection which requires a more detailed report. More time will also be required where the inspection is to cover any designated ruin in the churchyard.

In addition to fees, the professional adviser should be reimbursed for out-of-pocket expenses properly and reasonably incurred in connection with the inspection. VAT will be charged on the professional adviser's fees and expenses.

Preparing for the inspection

Well before the intended inspection, the professional adviser will write to the incumbent or PCC secretary to arrange a date. Various preparations are necessary so that the professional adviser can have access to all parts of the building, and complete the inspection in a safe manner.

Keys
All places normally kept locked, such as the tower, vestry, boiler house, storerooms etc., must be open for inspection or keys made available.

Ladders
Sturdy ladders must be provided for inspection of the roof timbers and roof coverings, and adequate labour to handle them. Ladders should reach up to roof plate level inside the building and to all parts outside. If ladders are not kept in the church, a builder should be asked to provide them. (The opportunity could be taken to clean out the gutters during the inspection, probably at extra cost.) The professional adviser cannot be expected to inspect any parts of the building that cannot be reached in safety.

Inventory, log book and other information
Churchwardens are required by law to maintain an inventory, listing all the church goods and a log book giving details of work carried out to the church. The log book should be available during the inspection, so that the professional adviser can note in the report brief details of repairs and other works carried out during the previous quinquennium. Details of servicing arrangements for items such as heating boilers and fire extinguishers should also be provided if requested, together with any archaeological reports. Details of any tree preservation orders or ruins designated by the Council for British Archaeology and the Royal Commission on the Historical Monuments of England should also be made available.

The inventory needs to be provided to check any items which the archdeacon has directed to be inspected, and to help the adviser to identify any other items that should be included.

Security

Care should be taken not to create any unsuitable security risks during the inspection. The local neighbourhood watch should be informed if such a scheme exists, and all doors and windows should be closed and checked by a responsible member of the church when the professional adviser has left.

Bells

The tower captain should be requested to ensure that the bells are down on the day of the inspection and to note any defects in the bellframe or difficulties in ringing experienced in recent years, especially any alterations in ringing characteristics.

The inspection and report

A model form of quinquennial inspection report is included in Appendix 3 as a guide to the amount of detail required. The CCC recommends it as the basis of future quinquennial inspection reports, to give greater coherence and clarity of presentation. A simple check list or 'tick box' approach is inadequate. The report needs to be of some substance to be of value to the PCC and to the other bodies that will see it.

Some advisers include approximate indications of cost, which are a great help to parishes in assessing likely expenditure. They should however be taken as only the most approximate of guides at this stage, given the limitations of the inspection.

Copies of the report must be sent to the incumbent, the secretary of the PCC, the archdeacon and the secretary of the DAC.

Professional advisers should always bear in mind that their reports will be read largely by non-professionals seeking guidance for future repairs. The report should be clear and action-orientated, with technical language kept to a minimum. It is helpful if the professional adviser attends a PCC meeting after the report has been submitted.

If the archdeacon finds that either a church or any designated articles have not been inspected, he can require the PCC to arrange for a supplementary inspection and recover the costs via the Diocesan Scheme.

Putting the work in hand

Planning

The quinquennial report is not a specification for repair work, and must not be used for this purpose. The scope and nature of necessary works summarised in the survey are extremely varied and the recommendations are intended only to help the PCC to establish repair priorities throughout the next five years. A comprehensive report will also offer guidance over a longer period than five years and should help to anticipate long-term repair liabilities, with associated cost implications. It may be cost effective to group together types of repair, small stonework items for example, to gain a cost advantage.

Before implementing any of the recommendations in the report, the professional adviser should be asked to suggest work that can be undertaken by local or volunteer labour and what work will require an experienced building contractor. Some work will probably need the attention of specialist conservators.

Much time is spent by PCCs discussing repairs and improvements, as well as raising funds to finance these works. Although responsibility in law for the care of the church rests with the PCC itself, it is sensible to appoint either a fabric officer or a committee, which may include people from the community who are not necessarily churchgoers, to deal with matters relating to the fabric. In this way, a repairs project can be seen as part of the church's mission to the community. The officer or committee must keep in close touch with the PCC, report to it frequently and remain subject to effective control by the PCC. The principal tasks of a fabric officer would be:

a   Liaising with the professional adviser at the time of the quinquennial inspection;

b   Taking a lead in implementing the report's recommendations, seeking grants from private and public sources, state aid, etc. as appropriate, in cooperation with the church treasurer.

c   Assisting the churchwardens in the annual inspection of the church, and in preparing the report on the fabric for the meeting of the PCC before the annual parochial church meeting.

d   Implementing a routine maintenance programme as set out in Appendix 6, in particular arranging for gutter clearance on a six- monthly basis.

The appointment of a press officer has been found extremely beneficial to ensure accurate reporting of progress in the local media. The opportunities given for mission and evangelism should not be overlooked (see *Mission in Mortar*, published by CCC 1993).

Funding the work

Although much of the financial burden for undertaking repairs will fall on the congregation, grants are available from a number of sources to assist in this respect.

English Heritage operates a church grant scheme to help with repairs to historic buildings in use for public worship, under section 3A of the Historic Buildings and Ancient Monuments Act 1953. However, statutory powers to make grants are limited, as are funds. English Heritage has to be very selective and generally gives priority to the most important churches in the greatest need. The criteria for making church grants are quite specific. A church must be of outstanding architectural and historic interest, listed grade I or II*, it must be in urgent need of major structural repairs and the cost of such repairs must be beyond the means of the congregation. Applications for grant aid should be made on the appropriate form signed by an authorised member of the church with day-to-day responsibility for repairs and countersigned by the archdeacon. The completed form should be submitted with photographs, copies of any professional reports (including the quinquennial inspection report) or specifications, copies of annual accounts and any other documents which may be relevant to the application, for example a current guidebook.

Applications are considered against the stated criteria and, if these are met, an English Heritage officer or commissioned consultant will visit the church and make an independent assessment on the nature, cost and urgency of all foreseeable repairs likely to be eligible for a grant. The application process may take up to six months to work through the system, and any offer will be subject to conditions. Guidance notes for applicants are available from English Heritage.

Other sources of grant aid available from private trusts are listed in *Sources of Grant Aid* available from the CCC. A full list of grant making trusts appears in *The Directory of Grant Making Trusts* published by the Charities Aid Foundation, and available in most public libraries. Local authorities are also empowered to make grants under the Local Authorities (Historic Buildings) Act 1962. The Historic Churches Preservation Trust and county historic churches trusts may be able to assist or advise of private charities which can help.

The setting up of a local church 'Building Preservation Trust' has many benefits in terms of fund raising and protection of funds given for building repairs.

Seeking approval
Faculty jurisdiction

Churches, their furnishings, and churchyards, are subject to the authority of the consistory court of the diocese, and no repair or alterations may be carried out, nor may any item of furnishing or other article (subject to some minor exceptions) be removed or acquired without its consent, which is given by way of a 'faculty'. The main legislation and its administration are set out in the Care of Churches and Ecclesiastical Jurisdiction Measure 1991, the Faculty Jurisidiction Rules 1992 and the Care of Churches and Ecclesiastical Jurisdiction Measure: Code of Practice issued in 1993. Every parish should possess their own copies (which can be obtained from Church House Bookshop). The archdeacon or the DAC secretary can help with any uncertainties or difficulties which arise; queries of a legal nature should be addressed to the diocesan registrar.

The consistory court receives advice on applications for faculty permission from the DAC. The DAC comprises archdeacons, clergy and laity as well as architects and surveyors, experts on archaeology, heating, bells and organs, and examines applications for faculty permission in great detail. The expertise of members of the DAC is extensive and their guidance, which is freely available, will be especially valuable to parishes before they make formal submissions.

A fee is payable when an application is made for a faculty. In some dioceses this is paid centrally by the Diocesan Board of Finance and the charge shared between parishes by way of the quota. Guidance on this will be available from the diocesan registrar. The application will be handled either by the archdeacon or by the chancellor, normally after obtaining the advice of the DAC.

Permissions from secular authorities

Approvals from other bodies may also be needed, and the professional adviser should be able to advise accordingly. These include English Heritage, if a grant is being applied for or has previously been given for earlier work, and English Nature if the building is a recognised bat roost.

Anglican churches in use for worship which are listed buildings are exempt from the need for local authority listed building consent although this does not apply to structures or objects fixed to the outside of the church or within its curtilage that are listed in their own right. This 'ecclesiastical exemption' is retained because of the Church's own system of care and control. It is therefore in everyone's best interests that the requirements of the faculty jurisdiction are obeyed.

Approval may however still be required from the local authority under the Building Regulations for building works including those involving new building, structural alterations or change of use, and from the local water board or National Rivers Authority if works involve drainage affecting water courses. In addition, any work which significantly affects the external appearance of the building, or changes its use constitutes 'development' within the meaning of the town and country planning legislation, and is not covered by the exemption. Planning permission in addition to faculty permission is therefore required for such works, which can include seemingly minor items such as a change in roof materials, or the provision of an oil storage tank.

Consulting other people and bodies

It will often be appropriate to consult other people and bodies in addition to those mentioned above; guidance on this is given in the Care of Churches and Ecclesiastical Jurisdiction Measure: Code of Practice.

Selecting the contractor

Having appointed the professional adviser, assessed budget costs, completed the specification, raised the funds and obtained the necessary approvals, competitive estimates should be sought for the repairs. It is always tempting to rely on the help of members of the church community or local tradesmen who may be able to undertake the work at apparently less cost than specialist contractors and without involving the professional adviser. However it is usually wiser in the long term to involve the professional adviser, especially where the church building is listed as being of special architectural or historic interest. Professional advice on repairs will ensure that they are well executed, and the PCC obtains good value for money.

The professional adviser will discuss the suitability of particular contractors to undertake specific types of repair and may recommend that specialist conservators be appointed to advise on important items, such as fittings, furnishings, wall paintings etc. General builders may be competent to carry out normal maintenance works but specialist contractors will be needed for the repair and replacement of masonry, repointing of stone and brickwork and repair of brickwork and lead. If the contractor is not well known, it will be sensible to obtain references, in particular of his past experience in working on similar church buildings.

Whilst small-scale repairs may be completed without the use of one of the standard forms of building contract, works which include several trades are more complex and require the discipline of such a procedure. It is also important that the obligations of the PCC and any appointed contractor are understood and agreed by both parties.

The professional adviser will suggest which of the standard forms of building contract should be used, and which of those issued by the Joint Contracts Tribunal (JCT) is the most appropriate for the particular project; an alternative type of contract may be recommended under very special circumstances. Most of the standard forms require a contract administrator, a role normally fulfilled by the professional adviser.

Larger repair schemes may also require the appointment of a project supervisor under the health and safety legislation.

Insurance

Insurance is not a substitute for risk management. It is better to avoid taking risks in the first place, for example by good security precautions against fire and theft, rather than to rely on insurance to put things right when a disaster happens.

Church inspection and repair

It is important that the PCC is aware of the risks that may arise with building work and of its own insurance obligations and those of any contractor employed. The PCC should also be aware of the need to take out insurance cover in respect of simple maintenance by members of the church community, relating to safety, accidental damage and negligence. Failure to appreciate the risks can give rise to unexpected claims, and the PCC's insurers should always be consulted in case of any doubt.

The liability for injury, loss or damage is defined by the building contracts. Under the forms of contract issued by the JCT for new buildings, a contractor is required to insure and indemnify the PCC against any claim due:

a) to personal injury or death of anybody caused by the works, except where due to the negligence of the PCC, its agent or employee or of a statutory undertaking;

b) to damage to property other than the works caused by the carrying out of the works and due to the contractor's negligence. The value of cover must be specified for this item;

c) to protect existing structures, their contents, the works and materials on site against 'specified perils', particularly fire.

Where works are to existing buildings or structures, the PCC will be required to insure against such risks in the joint names of the PCC and the contractor. Again, the PCC must seek the insurance company's assistance to ensure that adequate cover has been obtained.

Either party may be required to provide evidence that the appropriate insurances are in place before any works are commenced. The PCC must always ensure that all those working on the church have sufficient cover to meet the outcome of a claim.

Insurance cover for building works requires specialised advice to ensure that the interests of all parties to a contract are adequately protected. The parish should study JCT Practice Note 22 and consult their insurer.

Copies of specifications and invoices, which identify precisely what has been carried out, when and by whom, should be kept in the church log book together with any photographs which show details of the repairs and structure uncovered during the work. Accurate records of work can save time and money when considering future repairs and alterations and become over the years an historical archive of data on the building fabric. Records of archaeological finds and of inspections by archaeological consultants should also be made; records of 'no finds' are of equal importance. The PCC may be called upon by English Heritage or the DAC to employ the services of an archaeological consultant, either to record important elements of the standing structure or to investigate areas of the work exposed during the course of repairs, whether they be above or below ground. Copies of these records should also be retained in the log book.

Value added tax

Works of repair and maintenance to all buildings, including churches, are liable for VAT at the standard rate. This applies whether or not the building in question is listed. Such works include repair or renewal of floors, ceilings, walls, roofs and windows, and both external and internal redecoration.

It is possible to claim zero rating for an approved alteration to a listed secular building, defined as an alteration which requires listed building consent, and for which such consent has been given. However, in the case of a church such an alteration is defined as an approved alteration where the building is listed and is being used for ecclesiastical purposes, or would be used as such but for the works in question, and faculty approval has been obtained.

Responsibility for the collection of the tax rests with the contractor, who may be subject to penalties if his assessment differs from that of the local inspector. It is best to confirm the extent of liability with the local VAT office in writing before starting the contract. To zero-rate any of the work, the builder will need evidence that the church is listed, obtainable from the planning registry of the local authority and also a certificate of compliance from the church in the form prescribed by Customs and Excise.

Professional fees are always standard-rated, regardless of whether they relate to standard or zero-rated work.

The rates at which VAT is assessed are likely to change in future years, in line with the policy of the European Union. The current regulations are explained in a series of booklets available from Customs and Excise offices. Further general advice on VAT matters is contained in the leaflet *VAT and the Churches*, published by the Churches Main Committee. The local Customs and Excise officer may be able to assist in difficult cases.

appendix 1
Model diocesan scheme

The CCC and the Ecclesiastical Architects' and Surveyors' Association recommend this draft scheme to dioceses to enable them to carry out the requirements of the Inspection of Churches Measure 1955, as amended by the Care of Churches and Ecclesiastical Jurisdiction Measure 1991 ('the Measure').

1   This Scheme is established by the [name of diocese] Diocesan Synod by a resolution of [date]. It supersedes all previous schemes, and shall come into operation on [date].

2   The Scheme shall be administered through the diocesan [secretary] and all correspondence should be addressed to the secretary of the diocese.

3   All parish churches in the Diocese and all other consecrated churches and chapels and buildings licensed for public worship which the Measure requires to be inspected shall be inspected under this Scheme, together with articles, trees and ruins which are required to be inspected under the Measure; approximately one-fifth of the buildings are to be inspected each year.

4   PCCs should consult the Diocesan Advisory Committee about the choice of an inspecting architect/surveyor (professional adviser) if a change is necessary, and the appointment will be subject to the approval of the DAC.

5   The inspection of the church is to be visual, and such as can be made from ground level, ladders or accessible roofs, galleries or stagings. Parts of the structure which are inaccessible, enclosed, or covered will not be opened up unless specifically requested. The inspection is to include, so far as practicable, all features of the building, and to cover all aspects of conservation and repair. The PCC shall provide ladders and any other assistance as the professional adviser considers necessary.

6 From the notes taken at the inspection, the professional adviser shall prepare a report following the format set out in the current edition of *A Guide to Church Inspection and Repair*.

7 Within one calendar month from making the inspection, the professional adviser shall send a copy of the report to the archdeacon of the archdeaconry, the PCC of the parish in which the church is situated, the incumbent or priest in charge and to the secretary of the DAC.

8 The Diocesan Secretary shall be responsible for keeping a register of those buildings which are covered by the Scheme, and of appointed professional advisers. The register shall also include dates of inspections and reports.

9 The Diocesan Board of Finance shall establish a fund for and pay the fees for all inspections in the diocese. The fee shall be calculated taking account of the recommendations in the current edition of *A Guide to Church Inspection and Repair*, and reviewed each year by the Board following advice from the DAC.

10 Nothing in this Scheme shall alter the powers of an archdeacon to ensure the inspection of every church in his archdeaconry once in five years, as laid down in Sections 2 and 3 of the Inspection of Churches Measure 1955 as amended by the Care of Churches and Ecclesiastical Jurisdiction Measure 1991.

11 Any questions which may arise concerning the interpretation of this Scheme shall be referred to the Bishop's Council, the decision of which shall be binding.

12 This Scheme shall be subject to amendment only by means of a formal motion, presented after due notice to the diocesan synod, and approved by it.

appendix 2
Draft letter of appointment

From a new professional adviser to the secretary of the PCC

Dear

INSPECTION OF CHURCHES MEASURE 1955
(AS AMENDED)

[NAME OF CHURCH AND DIOCESE]

Thank you for inviting me to become professional adviser to your church, an appointment which I am happy to accept. I will undertake regular quinquennial inspections in accordance with the provisions of your diocesan scheme, and advise on, prepare specifications for, and oversee subsequent repairs where invited to do so.

I shall be pleased to accept this appointment in accordance with the following terms:

1. The inspection of the church will be visual, and such as can be made from ground level, ladders and any readily accessible roofs, galleries or stagings, and only selected areas will be examined in detail. Parts of the structure which are inaccessible, enclosed or covered will not normally be opened up unless specifically requested. The PCC shall provide ladders and any other necessary assistance. I shall be pleased to discuss my detailed requirements for these with you.

2. When I come to inspect, I will want to see the log book of alterations and reports, and the inventory of all articles. I will also need to see:-

Draft letter of appointment

    (a)   a list of any movable articles which the archdeacon has directed me to treat as of outstanding architectural, artistic, historical or archaeological value, or of significant monetary value, or at special risk of being stolen or damaged;

    (b)   details of any ruins in the churchyard (open or closed) which the CBA and the RCHME have designated as being of outstanding architectural, artistic, historical or archaeol-ogical value; and

    (c)   copies or details of any tree preservation orders affecting trees in the churchyard (open or closed).

3   The inspection will include as far as practicable all features of the building, covering all aspects of conservation and repair, and will include all articles, ruins and trees which the diocesan scheme requires to be covered.

4   The report will be prepared and presented to conform to the requirements of the diocesan scheme and set out in accordance with the recommendation of the current edition of *A Guide to Church Inspection and Repair*. The report will be submitted to the PCC, and a copy will be sent at the same time to the archdeacon, the incumbent, and the secretary of the DAC. Further copies will be issued in accordance with the directions contained in the scheme.

4   The report will not include a formal estimate of costs, nor a specification for repair works.

5   I shall deliver the report within twenty-eight days of carrying out the inspection.

6   My fee for the inspection and report will be [......(see note i)] [as laid down by the Diocese] [In addition to the fee charges mentioned in paragraph 3 above I shall charge the following out-of-pocket expenses:

    ....]

or

[My fees are inclusive of out-of-pocket expenses].

VAT at the standard rate will be charged additionally on all fees and expenses.

Appendix 2

          The conditions of appointment will be as set out in [the RIBA Standard Form of Agreement for the Appointment of an Architect (SFA/92)]/[the RICS Conditions of Engagement for Building Surveying Services] (copy attached).

7   The general good practice and spirit of our relationship will be as set out in *A Guide to Church Inspection and Repair*, published by Church House Publishing for the CCC. I suggest you obtain a copy of this if you do not already have one. Where I am required to write any subsequent specification and supervise, oversee or inspect repair work, this will be the subject of a separate agreement which will set out the scope of my services and the fee basis. (see note ii).

          I understand that my appointment as professional adviser will continue until terminated by either of us. Please keep me informed of any proposals or factors which may affect the care of your church, so that I may advise you to the best of my ability.

[In addition, as I am taking over this responsibility from another professional adviser, it would be helpful if I could have his/her name and address so that I may write to him/her. Please send me copies of previous reports or at least of the most recent].

Please confirm that these terms of appointment are acceptable to you and your PCC. I attach a duplicate copy of the letter for you to sign and return to me as a record of the agreed appointment.

I am sending a copy of this letter to the incumbent, for information.

Yours sincerely

Professional Adviser

Notes:

(i) Fees for the inspection and report may be charged on a time or lump sum basis subject to the Diocesan Scheme. Where a time charge is to be made, the hourly rate should be stated and an estim-ate of the number of hours likely to be taken should be provided.

(ii) Any subsequent specification and inspection of repair work should be the subject of a separate appointment. The fees for such work will normally be charged on a time or percentage basis.

(iii) Fees may be paid by the Diocese direct or by the PCC which may then receive re-imbursement depending on the scheme currently in force in the Diocese.

appendix 3
Outline quinquennial inspection report

Preliminary information Name of church, diocese, archdeaconry.

Name of professional adviser, address and telephone number.

Date of inspection and report, date of previous inspections, record of weather conditions.

Key plan, with a drawn scale where possible.

Brief description of the building, including orientation.

First report only:
List the trees in the churchyard, noting any subject to tree preservation orders, note whether the church is within a conservation area, the church's historical background, brief architectural history, materials used in the construction, seating capacity, site access, provision for disabled people, parking facilities.

Main Report
Limitations

State limitations of the report:

Whether it is made from the ground or from other accessible floor levels, ladders and readily accessible locations.

That the inspections are visual. Opening up of enclosed spaces is excluded, even if further inspection of these spaces may be recommended.

If appropriate, list items not inspected.

Note that the report is restricted to general condition of the building and its defects.

Quinquennial inspection report

1	SCHEDULE OF WORKS COMPLETED SINCE THE PREVIOUS QUINQUENNIAL REPORT	List repairs carried out since the last inspection including items of emergency repair. List alterations, additions and demolitions.
2	GENERAL CONDITION	Describe the general condition of the building noting any particular movements, subsidence and settlement, areas of damp penetration, or general areas of damage and decay. Note any particular work undertaken outside the churchyard which might have an impact on the church and its setting.

EXTERNAL

3	ROOF COVERINGS	Systematically record materials, construction, general condition, including ridges, hips, valleys, parapet wall gutters, cess boxes, chutes, flashings, and any special features.
4	RAINWATER GOODS AND DISPOSAL SYSTEMS	Record materials, condition and cleanliness, assess whether adequate.
5	BELOW GROUND DRAINAGE	Comment on storm drains, soakaways, foul drains, inspection chambers and rodding eyes and their condition.
6	PARAPETS AND UPSTAND WALLS	Construction and condition of parapets, copings, cappings, finials, crosses.
7	WALLING	Record materials and general condition of all walling to towers and spires, walls, crossing walls, referring to buttresses, to stonework details such as cills, mullions, stringcourses, arches, lintels, carved and moulded features. Plinths, gratings, air bricks. Note the condition of pointing.

Appendix 3

8	TIMBER PORCHES, DOORS AND CANOPIES	Comment on the materials and genera condition of all timber structures, including doors and their frames, timber and metal window frames, commenting on external finishes.
9	WINDOWS	Comment on the condition of external window openings, stonework, saddlebars, glazing, including the leading, condensation trays and ferramenta.

INTERNAL

10	TOWERS, SPIRES	Comment on the condition of the tower internal walling and spire from nearest accessible point internally.
		Note general condition of bells and bellframe, headstocks and rope guides.
		Timber floors, supporting structures, noting any beam ends which need further investigation
		Louvres and bird mesh.
		Access provision, ladders, trapdoors etc.
11	CLOCKS AND THEIR ENCLOSURES	Note general condition of external enclosures, any evidence of routine maintenance, general information on condition.
12	ROOF AND CEILING VOIDS	Where accessible, note general condition Include signs of water penetration, structural failure, rot and insect attack.
		Where suspended ceilings exist, comment on materials and general condition.
		Where possibility of asbestos exists, its condition and implications for removal.
13	ROOF STRUCTURES, CEILINGS, CEILURES	Comment on materials and general condition of all exposed elements.
		Include braces, fixing methods, decorative panels.

Quinquennial inspection report

14	UPPER FLOORS, BALCONIES, ACCESS STAIRWAYS	Comment on the construction and condition of upper floors within the main building. Note requirements for ventilation. Report on general condition of balconies, stairways and balustradings, noting any particular areas needing improvemen under health and safety legislation.
15	PARTITIONS, SCREENS, PANELLING, DOORS AND DOOR FURNITURE	Comment on materials and general condition of all screens, panelling, partitions, doors, frames and ironmongery. Comment on any carved items, painted panels and other items of particular merit.
16	GROUND FLOOR STRUCTURE, TIMBER PLATFORMS	Comment on materials and general condition, ventilation and adequacy. Report on general conditions of timber platforms, pew platforms.
17	INTERNAL FINISHES	Comment on materials and condition of wall and ceiling finishes. Note dampness, areas of decayed plaster and any other apparent defects.
17	INTERNAL FINISHES	Comment on materials and condition of wall and ceiling finishes. Note dampness, areas of decayed plaster and any other apparent defects.
18	FITTINGS, FIXTURES, FURNITURE AND MOVABLE ARTICLES	Comment on condition of important fittings, fixtures and movable articles. Note particularly any designated by the archdeacon for inspection, in a separate letter. Note defects and make recommendations for improving security (in a separate letter). Note whether conservation or other specialist advice is required.

Appendix 3

19	TOILETS, KITCHENS, VESTRIES, ETC.	General condition, fitness for purpose, cleanliness.
20	ORGANS AND OTHER MUSICAL INSTRUMENTS	Comment on general condition and access provision.
21	MONUMENTS, TOMBS, PLAQUES, ETC.	Comment briefly on condition and make recommendations for specialist advice where **necessary.**
22	SERVICE INSTALLATIONS GENERALLY	Note that the report and comments are based on a visual examination only and that no tests of services have been **undertaken.**
		Make recommendations for testing, as appropriate.
23	HEATING INSTALLATION	State type of system installed, fuel, age, apparent condition and existence of maintenance agreements (PCC to advise).
24	ELECTRICAL INSTALLATION	Note location, apparent condition of incoming mains, meters and distribution boards.
		Note when last inspected by NICEIC contractor (PCC to advise).
25	SOUND SYSTEM	Comment on the provision and condition of sound systems, loop systems, whether regularly maintained under a maintenance agreement.

Quinquennial inspection report

26 LIGHTNING CONDUCTOR

Comment on condition, when last inspected; make recommendations for testing and improvement in accordance with the latest British Standard.

27 FIRE PRECAUTIONS

Note number, position and types of fire extinguishers provided.

Examine records of maintenance for appliances.

28 DISABLED PROVISION AND ACCESS

Comment on provision for the disabled, including access to various parts of the church and recommendations for necessary improvements.

29 SAFETY

Comment in general on the safety of the church for its users and visitors.

30 BATS

Comment on any known locations of bats, reports known from any bat groups, likely bat roosts, and implications for future repairs.

CURTILAGE

31 CHURCHYARD

Comment on general condition of the grassed and planted areas.

32 RUINS

Inspect and comment on any ruin in the churchyard, noting any known to be designated as being of outstanding architectural, artistic, historical or archaeological value (PCC to advise).

33 MONUMENTS, TOMBS AND VAULTS

Comment on general condition, making specific reference to any obvious defects.

34 BOUNDARY WALLS, LYCHGATES AND FENCING

Briefly describe in general terms materials and condition of all elements.

35 TREES AND SHRUBS

Note any trees or shrubs likely to injure persons or damage the fabric of the building.

36 HARDSTANDING AREAS

Comment on general condition of paths, pavings, hardstandings, steps, car parking areas and surface water drainage.

Appendix 3

37 MISCELLANEOUS   Comment on garden sheds and other site features not mentioned above, rubbish disposal etc.

38 LOG BOOK   Inspect the log book provided by the PCC.

Comment on requirement for reports from the fire prevention officer, crime prevention officer, security consultant, insurers, etc.

Quinquennial inspection report

Recommendations

List items under the following degrees of priority, where possible with broad budget costs. Note items that might safely be entrusted to unskilled labour and others which may qualify for grant aid.

Note specifically the following:

1   Urgent works requiring immediate attention.

2   Works recommended to be carried out during the next twelve months.

3   Works recommended to be carried out during the quinquennial period.

4   Works needing consideration beyond the quinquennial period.

5   Works required to improve energy efficiency of the structure and services.

6   Works required to improve disabled access.

Draft form of 'Explanatory Notes' to be added to all inspection reports

A   Any electrical installation should be tested at least every quinquennium by a registered NICEIC electrician, and a resistance and earth continuity test should be obtained on all circuits. The engineer's test report should be kept with the church log book. This present report is based upon a visual inspection of the main switchboard and of certain sections of the wiring selected at random, without the use of instruments.

B   Any lightning conductor should be tested every quinquennium in accordance with the current British Standard by a competent engineer, and the record of the test results and conditions should be kept with the church log book.

C   A proper examination and test should be made of the heating apparatus by a qualified engineer, each summer before the heating season begins.

D   A minimum of two water type fire extinguishers (sited adjacent to each exit) should be provided plus additional special extinguishers for the organ and boiler house, as detailed below.

Appendix 3

Large churches will require more extinguishers. As a general rule of thumb, one water extinguisher should be provided for every 250 square metres of floor area.

Summary:

Location	Type of Extinguisher
General area	Water
Organ	CO_2
Boiler House	
Solid fuel boiler	Water
Gas fired boiler	Dry powder
Oil fired boiler	Foam (or dry powder if electricity supply to boiler room cannot easily be isolated).

All extinguishers should be inspected annually by a competent engineer to ensure they are in good working order.

Further advice can be obtained from the fire prevention officer of the local fire brigade and from your insurers.

E   This is a summary report only, as it is required by the Inspection of Churches Measure; it is not a specification for the execution of the work and must not be used as such.

The professional adviser is willing to advise the PCC on implementing the recommendations, and will if so requested prepare a specification, seek tenders and oversee the repairs.

F   Although the Measure requires the church to be inspected every five years, it should be realised that serious trouble may develop in between these surveys if minor defects are left unattended. Churchwardens are required by the Care of Churches and Ecclesiastical Jurisdiction Measure 1991 to make an annual inspection of the fabric and furnishings of the church, and to prepare a report for consideration by the meeting of the PCC before the Annual Parochial Church Meeting. This then must be presented with any amendments made by the PCC, to the Annual Parochial Church Meeting. **The PCC are strongly advised to enter into contract with a local builder for the cleaning-out of gutters and downpipes twice a year.**

The Churchwarden's Year gives general guidance on routine inspections and house keeping, and general guidance on cleaning is given in *Handle with Prayer*, both published for the CCC by Church House Publishing.

G   The PCC are reminded that insurance cover should be index-linked, so that adequate cover is maintained against inflation of building costs. Contact should be made with the insurance company to ensure that insurance cover is adequate.

H   The repairs recommended in the report will (with the exception of some minor maintenance items) be subject to the faculty jurisdiction.

I   Woodwork or other parts of the building that are covered, unexposed or inaccessible have not been inspected. The adviser cannot therefore report that any such part of the building is free from defect.

This appendix is based on *A Guide for the Quinquennial Inspection of Churches*, Diocese of Birmingham, 1993.

appendix 4
Advice to DACs on assessing potential professional advisors

When a PCC applies to a DAC for approval to the appointment of a professional adviser, the DAC will wish to examine closely the experience of the applicant particularly in relation to the repair of historic buildings. It will also bear in mind the desirability of bringing new people into the field of church restoration. The following criteria might be used when assessing applicants:

- For which churches does the applicant already act as professional adviser?

- Has the applicant specified and overseen repairs to secular listed buildings?

- Has the applicant prepared specifications for which grant aid was offered by English Heritage?

- How does the applicant assess fees?

- What is the level of the applicant's professional indemnity insurance?

- Is the applicant a sole practitioner? If not, what back-up staff are available?

Where the candidate is not known to the Committee, the DAC should request a copy of previous quinquennial inspection reports or a similar survey on a secular building.

Evidence of membership of societies such as the Royal Society of Arts (FRSA) or the Association for Studies in the Conservation of Historic Buildings (ASCHB), both of which receive members by invitation only, and of the Ecclesiastical Architects' and Surveyors' Association (EASA), which has strict entry qualifications, may indicate a certain level of awareness and understanding.

Attendance at short courses run by the Institute of Advanced Architectural Studies at York University, the Society for the Protection of Ancient Buildings, as part of the professional adviser's continuing professional development, are also useful indicators of knowledge and awareness as are diplomas in conservation studies.

The Royal Institution of Chartered Surveyors maintains a list of chartered surveyors accredited as having acceptable experience in the conservation of historic buildings.

The Royal Institute of British Architects offers a nomination service, and can provide details of architects having particular experience in historic buildings conservation.

appendix 5

Churchwardens' responsibilities under the Care of Churches and Ecclesiastical Jurisdiction Measure 1991

1. * Compile and maintain a full terrier of all lands belonging to the church and an inventory of all articles belonging to the church.
   Section 4 (1) (a)

2. * Maintain a log book containing a full note of all work done to the church with relevant details and papers, including a note of the location of all relevant documents not kept with the log book.
   Section 4 (1) (b)

3. Send a copy of the inventory and notice of any alterations to the person designated by the bishop.
   Section 4 (4)

4. * Inspect the fabric of the church at least annually.
   Section 5 (1) (a) + (7)

5. * Make an annual report on the fabric to the PCC at its meeting next before the annual meeting.
   Section 5 (1) (b) and (3)

6. * Make the annual report on the fabric, with any amendments made by the PCC, to the annual parochial church meeting, on behalf of the PCC.
   Section 5(1) (b) and (3)

7. Produce to the PCC as soon as possible after the beginning of the year, the terrier, inventory and log book, brought up to date, together with any other documents which the churchwardens consider likely to assist the PCC to discharge its functions in relation to the fabric of the

church and the articles appertaining to it, and a signed statement that the contents of the terrier, inventory and log book are accurate.

Section 5(4)

8. Comply with any order by the archdeacon to remove articles to places of safety.

Section 21

* To be carried out in consultation with the minister (sections 4(2) and 5(2))

appendix 6
A guide to the routine maintenance and inspection of church property

It is good practice for the PCC to appoint a fabric officer to take care of the routine maintenance of the church. This officer must report to the PCC and remain subject to its control and direction. The Care of Churches and Ecclesiastical Jurisdiction Measure 1991 requires the churchwardens to inspect the fabric of the church at least once a year, to produce a report on the fabric of the church and the articles belonging to it to the PCC, and to make that report to the annual parochial church meeting on behalf of the PCC. The following list gives an indication of the time of year when certain jobs should be done. It is not exhaustive.

	Whenever necessary inspect gutters and roofs from ground level and inside especially when it is raining.
	Clear snow from vulnerable areas.
	Clear concealed valley gutters.
Spring, early summer	Make full inspection of the church for annual meeting.
	Check church inventory and update log book.
	Check bird-proofing to meshed openings.
	Sweep out any high level spaces. Check for bats and report any finds to English Nature.
	Cut any ivy starting to grow up walls and poison.

Routine maintenance and inspection

Spray around the base of the walls to discourage weed growth.

Check heating apparatus and clean flues.

Arrange for routine servicing of heating equipment.

Check interior between second week of April and second week of June for active beetle infestation and report findings to the professional adviser.

Check all ventilators in the floor and elsewhere and clean out as necessary.

Spring clean the church.

Summer

Cut any church grass.

Cut ivy growth and spray (again).

Re-check heating installation before autumn and test run.

Arrange for any external painting required.

Autumn

Check gutters, downpipes, gullies, roofs etc. after leaf fall.

Rod out any drain runs to ensure water clears easily, especially under pavements.

Inspect roofs with binoculars from ground level, counting number of slipped slates, etc. for repair.

Clean rubbish from ventilation holes inside and out.

Check heating installation, lagging to hot water pipes etc. and repair as necessary.

Appendix 6

Winter	Check roof spaces and under floors for vermin and poison.
	Check under valley gutters after cold spells for signs of leaking roofs.
	Bleed radiators and undertake routine maintenance to heating systems.
	Check temperature in different areas of the building to ensure even temperature throughout and note any discrepancies.
Annually	Arrange for servicing of fire extinguishers.
	Inspect abutting buildings to ensure there is no build up of leaves or other debris against the walls.
	Check the condition of outside walls, windows, sash cords, steps and any other areas likely to be a hazard to people entering the building.
	Check the extent of any insurance cover and update as necessary.
Every 5 years	Arrange for testing of the electrical systems.
	Arrange for the testing of any lightning protection.

It is vital, especially with older buildings, to keep them warm and well ventilated at all times. The fabric officer should ensure that such ventilation is taking place, especially after services.

appendix 7

A guide to getting the work started

Instruct PA to undertake QIR

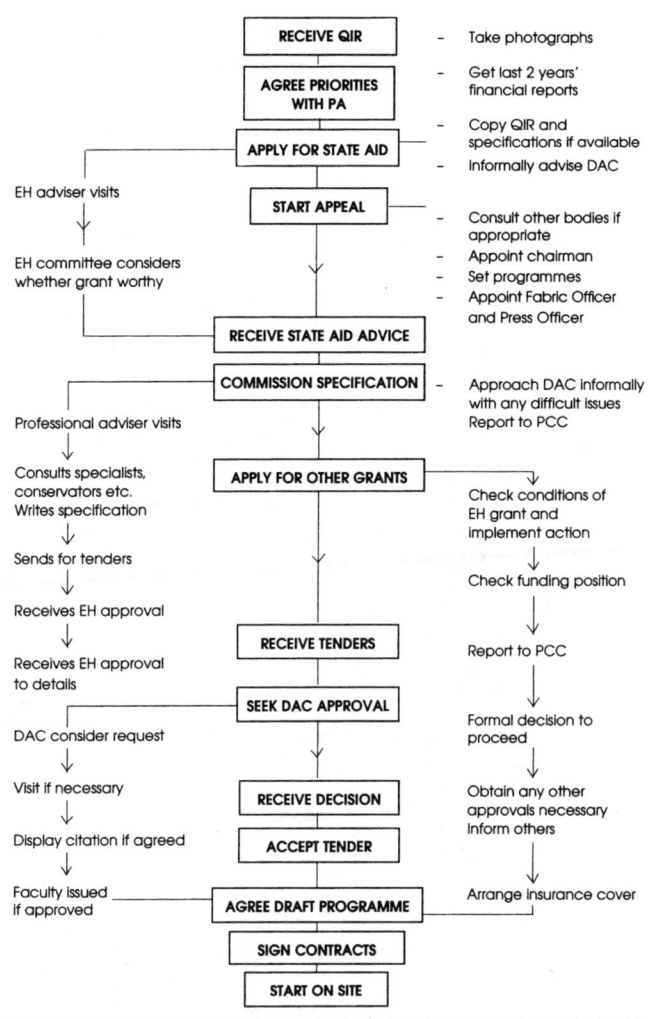

```
                              ┌─────────────────┐
                              │  RECEIVE QIR    │──  Take photographs
                              └─────────────────┘
                                       ↓          ─  Get last 2 years'
                              ┌─────────────────┐    financial reports
                              │ AGREE PRIORITIES│
                              │     WITH PA     │  ─  Copy QIR and
                              └─────────────────┘    specifications if available
                                       ↓
                              ┌─────────────────┐  ─  Informally advise DAC
            ┌─────────────────│ APPLY FOR STATE AID
            │                 └─────────────────┘
  EH adviser visits                    ↓
            ↓                 ┌─────────────────┐
            Y                 │  START APPEAL   │  ─  Consult other bodies if
            ↓                 └─────────────────┘     appropriate
  EH committee considers               ↓          ─  Appoint chairman
  whether grant worthy                            ─  Set programmes
            │                                     ─  Appoint Fabric Officer
            └─────────────────┐                      and Press Officer
                              ┌─────────────────┐
                              │RECEIVE STATE AID ADVICE│
                              └─────────────────┘
                                       ↓
            ┌─────────────────┤COMMISSION SPECIFICATION│─  Approach DAC informally
            │                 └─────────────────┘        with any difficult issues
 Professional adviser visits          ↓                   Report to PCC
            ↓
 Consults specialists,        ┌─────────────────┐
 conservators etc.            │APPLY FOR OTHER GRANTS│     ↓
 Writes specification         └─────────────────┘      Check conditions of
            ↓                          ↓               EH grant and
 Sends for tenders                                     implement action
            ↓                                              ↓
 Receives EH approval                                  Check funding position
            ↓                 ┌─────────────────┐          ↓
                              │ RECEIVE TENDERS │      Report to PCC
 Receives EH approval         └─────────────────┘          ↓
 to details                            ↓
                              ┌─────────────────┐      Formal decision to
 DAC consider request         │ SEEK DAC APPROVAL│     proceed
            ↓                 └─────────────────┘          ↓
 Visit if necessary                    ↓
            ↓                 ┌─────────────────┐      Obtain any other
                              │ RECEIVE DECISION│      approvals necessary
 Display citation if agreed   └─────────────────┘      Inform others
            ↓                          ↓                   ↓
                              ┌─────────────────┐
 Faculty issued               │  ACCEPT TENDER  │      Arrange insurance cover
 if approved                  └─────────────────┘
            │                          ↓
            └─────────────────┤AGREE DRAFT PROGRAMME│
                              └─────────────────┘
                                       ↓
                              ┌─────────────────┐
                              │ SIGN CONTRACTS  │
                              └─────────────────┘
                                       ↓
                              ┌─────────────────┐
                              │  START ON SITE  │
                              └─────────────────┘
```

The left hand column indicates work necessary by others, the central column the critical order for action by the PCC, and the right hand column shows supplementary actions and decisions necessary.

appendix 8
A guide to the faculty procedure

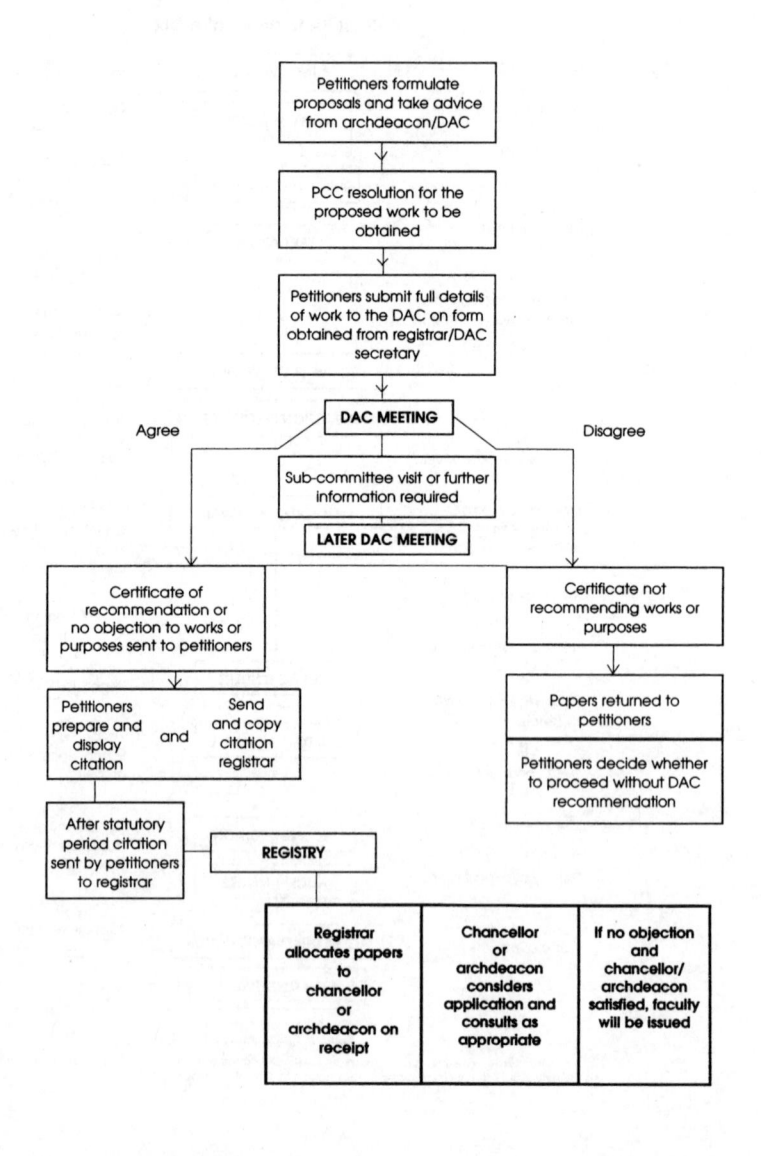

```
┌─────────────────────────────┐
│ Petitioners formulate       │
│ proposals and take advice   │
│ from archdeacon/DAC         │
└─────────────────────────────┘
              ↓
┌─────────────────────────────┐
│ PCC resolution for the      │
│ proposed work to be         │
│ obtained                    │
└─────────────────────────────┘
              ↓
┌─────────────────────────────┐
│ Petitioners submit full     │
│ details of work to the DAC  │
│ on form obtained from       │
│ registrar/DAC secretary     │
└─────────────────────────────┘
              ↓
         ┌──────────────┐
  Agree  │ DAC MEETING  │  Disagree
 ←───────└──────────────┘───────→
              ↓
┌─────────────────────────────┐
│ Sub-committee visit or      │
│ further information required│
└─────────────────────────────┘
      ┌──────────────────┐
      │ LATER DAC MEETING│
      └──────────────────┘
```

Agree branch:
- Certificate of recommendation or no objection to works or purposes sent to petitioners
- Petitioners prepare and display citation **and** Send and copy citation registrar
- After statutory period citation sent by petitioners to registrar

Disagree branch:
- Certificate not recommending works or purposes
- Papers returned to petitioners
- Petitioners decide whether to proceed without DAC recommendation

REGISTRY
- Registrar allocates papers to chancellor or archdeacon on receipt
- Chancellor or archdeacon considers application and consults as appropriate
- If no objection and chancellor/archdeacon satisfied, faculty will be issued

appendix 9
Further reading

The Archbishops' Council	*Care of Churches and Ecclesiastical Jurisdiction Measure: Code of Practice* (1993), CHP
CCC	*The Care of Church Plate*, CHP
CCC	*Church Log Book*, CHP
CCC	*Church Property Register (Terrier and Inventory)*, CHP
CCC	*The Churchwarden's Year: a Calendar of Church Maintenance,* illustrated by Graham Jeffery, CHP
CCC	*Fundraising for your Church Building*, CHP
CCC	*How to Look after your Church*, CHP
CCC	*Looking after your Church* (video), CHP
CCC	*Mission in Mortar*
P. Cunningham	*How old is that Church?* (revised edition, 1994), Marston House
	Care of Churches and Ecclesiastical Jurisdiction Measure 1991, The Stationery Office
	Faculty Jurisdiction Rules 1992, The Stationery Office
	Faculty Jurisdiction of the Church of England (Second edition, 1993) Newsom Sweet & Maxwell

CCC	–	Council for the Care of Churches
CHP	–	Church House Publishing

appendix 10
Useful addresses

Church bodies
Council for the Care of Churches, Church House, Great Smith Street, London SW1P 3NZ
(Tel: 020 7898 1866)
Churches Main Committee, 1 Millbank, London SW1P 3JZ*
(Tel: 020 7898 1861)
Church Commissioners, 1 Millbank, London SW1P 3JZ*
(Tel: 020 7898 1000)
Liturgical Commission, Church House, Great Smith Street, London SW1P 3NZ
(Tel: 020 7898 1364)
Church House Publishing, Church House, Great Smith Street, London SW1P 3NZ
(Tel: 020 7898 1000)

Amenity societies
The Society for the Protection of Ancient Buildings, 37 Spital Square, London E1 6DY
(Tel: 020 7377 1644)
The Twentieth Century Society, 70 Cowcross Street, London EC1M 6EJ
(Tel: 020 7250 3857)
The Victorian Society, 1 Priory Gardens, London W4 1TT
(Tel: 020 8994 1019)
The Ancient Monuments Society, St Ann's Vestry Hall, 2 Church Entry, London EC4V 5HB
(Tel: 020 7236 3934)

* Elizabeth House, 39 York Road, London SE1 7NQ (until December 2001)

Useful addresses

The Georgian Group, 6 Fitzroy Square, London W1P 6DX
(Tel: 020 7387 1720)

The Council for British Archaeology, Bowes Morrell House, 111 Walmgate, York YO1 9WA
(Tel: 01904 671417)

Joint Committee of National Amenity Societies, St Ann's Vestry Hall, 2 Church Entry, London EC4V 5HB
(Tel: 020 7236 3934)

Professional associations

Association for Studies in the Conservation of Historic Buildings (ASCHB), Institute of Archaeology, 31-34 Gordon Square, London WC1H 0PY
(Tel: 020 7973 3326)

Ecclesiastical Architects' and Surveyors' Association (EASA), Property Dept, Diocese of Salisbury, Church House, Crane Street, Salisbury SP1 2QB
(Tel: 01722 411 933)

Royal Institute of British Architects, 66 Portland Place, London W1B 1AD
(Tel: 020 7580 5533)

Royal Institution of Chartered Surveyors, 12 Great George Street, London SW1P 3AD (Tel: 020 7222 7000)

Others

Charities Aid Foundation, Kings Hill, West Malling, Kent ME19 4TA
(Tel: 01732 520000)

Department of Culture, Media and Sport, 2–4 Cockspur Street, London, SW1Y 5DH
(Tel: 020 7211 6200)

Appendix 10

English Heritage, Fortress House, 23 Savile Row,
London W1X 1AB
(Tel: 020 7973 3000)

English Nature, Northminster House, Peterborough
PE1 1UA
(Tel: 01733 455000)

National Association of Decorative and Fine Arts Societies,
8 Guildford Street, London WC1N 1DA
(Tel: 020 7430 0730)

York University, Institute of Advanced Architectural Studies,
Kings Manor, York YO1 2EP
(Tel: 01904 433 901)

Central Council of Church Bell Ringers, c/o Penmark House,
Woodbridge Meadows, Guildford, Surrey GU1 1BL